All living organisms other than plants, bacteria and viruses are classified as animals. But one often hears people talking about animals and birds, or animals and fishes, when they are really referring to mammals and these other groups. It is the mammals that form the subject of this book.

Mammals have their roots in the mammal-like reptiles, or Therapsida, a group which first appeared in late Carboniferous times about 280 million years ago. But it was not until late Triassic or early Jurassic times, about 190 million years ago, that the first animals that we would recognize as mammals came on the scene. During the Cretaceous period, from 136 million to 65 million years ago, the evolution of the mammals continued—although fossils are relatively rare—and from that time onwards they have diversified into all the forms that we know today.

Many groups of mammals have become extinct during this period of time and, as we shall see later, others are in the process of disappearing at the present time. The living members of the Class Mammalia are divided into three sub-classes: the Prototheria which contains the single order Monotremata and includes the duck-billed platypus and the echidna; the Metatheria—again with a single order, the Marsupialia—comprising, among other forms, the kangaroos and wallabies, the koala and the possums; and the Eutheria, containing 18 orders, and including groups such as the rodents, bats, carnivores, whales, seals and ungulates.

Mammals are generally considered to be the highest animals in the evolutionary scale, but this may be at least partly because man himself is a mammal. Certainly they are among the most complex animals anatomically and physiologically, only equalled in some respects by the birds. However, they are considerably more advanced than the birds in the development of intelligence, in the manner in which they protect their young during embryonic and foetal life, and in the more extended period of parental care and training which the young receive after birth. They are undoubtedly one of the most varied groups of animals, ranging from those that live on the land to those that use the sea as their environment; and, of course, some fly. Because of this diversity there are only a few characters which they all share in common.

By definition, a mammal is an animal in which the young are nourished, for a shorter or longer period after birth, on milk secreted by the mother. Mammals are also characterised by their possession of a protective and insulating hairy covering on the surface of the body, although this coat may be greatly reduced in some forms, for example the whales and manatees. It is true that some birds have hair-like structures among their plumage, but these are quite different in origin, being formed from modified feathers.

Internally there are a number of features which distinguish mammals from other groups. For instance, the lower jaw of a mammal is formed of a single bone, the dentary, whereas in other forms the lower jaw is constructed of several bones. In mammals these other bones become part of the structure of the ear. Mammals alone possess a diaphragm separating the chest cavity from the abdominal cavity. The brain too differs; that part known as the fore-brain is relatively much larger in mammals and because of its greater surface area it is convoluted or wrinkled. It is this part that is responsible for the increased intelligence of mammals.

An immensely important feature, which mammals share with birds, is that they are warm-blooded. They are usually active animals; their basic body chemistry, or metabolism, operates at a relatively high rate and this, in conjunction with their insulating layer of hair and a mechanism in the brain for keeping the body temperature remarkably uniform, enables them to maintain themselves under a very wide range of climatic conditions.

The three sub-classes of mammals did not lead on from one to another. The prototherians probably arose from a long-extinct group of mammals, the docodonts, in the Jurassic. Perhaps in common with all the early mammals, they lay eggs which are deposited in a pouch in the case of the echidna, but

WILD ANIMALS

A picture survey

(Originally published as *A Source Book of Wild Animals*)

Warren Phelops

WILD ANIMALS
A picture survey

(Originally published as *A Source Book of Wild Animals*)

Compiled by J. Lucas

A PICCOLO BOOK

PAN BOOKS LTD · LONDON

First published in Great Britain 1971 by Ward Lock Ltd.
This edition published 1973 by Pan Books Ltd.,
33 Tothill Street, London S.W.1

© WARD LOCK LIMITED 1971

ISBN 0 330 23670 9

Also available in the same series
MOTOR-CARS: A picture history
AIRCRAFT: A picture history
SHIPS: A picture history
LOCOMOTIVES: A picture history

Printed Offset Litho in Great Britain by
Cox & Wyman Ltd., London, Fakenham and Reading

are laid in a burrow by the platypus. In the echidna the ducts from the two milk glands open within the pouch, and the young lick or suck the milk as it is ejected. Young platypuses and echidnas are very poorly developed at hatching and are hidden until they are strong enough to venture out on their own.

The marsupials and the placental mammals probably sprang from another order of mammals, the pantotheres, in the mid-Cretaceous, but have since evolved along separate lines. The marsupials may, at first, have been widely distributed throughout the world, though fossils are all too rare for this to be certain. However, at the beginning of Eocene times the placental mammals evolved very rapidly and being more efficient and better adapted than the marsupials, soon brought about their complete extermination. That is, except in two regions, where they still live: Australia and South America. Australia became isolated from the Asiatic landmass at about the end of the Cretaceous, and South America similarly became isolated from North America at the beginning of the Eocene. Australia's isolation occurred before the placental mammals invaded the region, and hence the marsupials continued to evolve without competition. South America, on the other hand, was cut off after some primitive placentals had evolved, so that they and the marsupials existed side by side, but no advanced placentals were able to compete until the land bridge was re-established much later. When this occurred, many species of marsupials were exterminated and only a remnant is found there now.

The essential feature of the marsupials is their method of reproduction. The young are born more or less as larval animals after a short gestation period, and then make their way to the mother's pouch. Here they become attached to the nipples, which are located within the pouch. They remain attached for some time being fed with milk, which has to be pumped into them because, at least in the early stages, they are too small and weak to suck. They gradually grow, at first leaving the pouch for brief periods, until they are developed enough for independent life.

In their evolution, marsupials have paralleled to an astonishing degree the range of forms known among placental mammals. In some instances the physical appearance is very similar; thus the carnivorous thylacines can be compared with the wolves of other regions, and the small dasyures or 'native cats' with small cats, weasels and martens. In other instances it is the ecological similarity which is so striking, as with the kangaroos and wallabies which have a place in the web of life very like that of deer, antelopes and gazelles. However, although there are gliding marsupials, no forms have taken to the air in the same way as bats, and there are no marine forms paralleling whales and seals.

At present, the placental mammals, or Eutheria, are the dominant mammals throughout the world. One reason for this is certainly the care which they lavish on their offspring. The young go through a very considerable period of prenatal growth within the mother's uterus and are born at an advanced stage of development. During the gestation period they are attached to the mother by the placenta, through which food materials and oxygen pass from the mother to the growing embryo. The degree of development at birth varies widely. Some young, like those of rodents, carnivores and man are still helpless and need a shorter or longer period of care after birth, but the young of many species of ungulates and of whales are able to follow their mothers within an hour or two.

I said earlier that many groups of mammals had arisen, flourished briefly and then become extinct. In Jurassic times, roughly 190 million to 136 million years ago, there were five well-established orders of mammals, but all are now extinct. At least nine other orders have become extinct since then. During late Cretaceous and early Eocene times, parts of Africa, at least, were populated with giant oxen, giant sheep and other very large forms—including a rhinoceros which grew to a length of 25 ft and was 18 ft tall at the shoulder. These species have long since disappeared, and their descendants, including the modern rhinoceros, hippopotamus and giraffe, are really old-fashioned relics, doomed eventually to disappear in the natural course of events.

This is the central point of the present concern for conservation, for it is recognized that evolution is not a thing of the past, it is still continuing, although in most instances imperceptibly. Man's time scale is too short to be able to appreciate evolutionary changes as they are taking place. The present approach of conservation is not to halt evolution and turn the world into a museum—this would be impossible—but to prevent mankind from unnecessarily causing the extinction of species. The point is that if man causes the disappearance of species, these are not replaced by forms which are better adapted to their environment, whereas during the process of evolution new forms are produced which survive if they are an 'improved model', or become extinct in their turn if they are not.

Since the year 1600, at which time we first started to get descriptions of animals that we can now recognize, 36 species of mammals have become extinct. Of these perhaps one quarter disappeared as a result of natural causes, and the remainder through human interference. By far the most important cause was over-hunting but destruction or disturbance of the animal's habitat was almost equally disastrous.

At least 275 species and races of mammals are in considerable danger of becoming extinct, and of these only about 14 per cent are threatened through natural causes. The remainder, 86 per cent, are endangered through man's stupidity, thoughtlessness or greed. The mammals involved are not necessarily competing with man for living space for himself, his domestic animals or his crops, though some are. They range in size from the volcano rabbit of Mexico weighing a pound or two to the blue whale, the largest animal ever to have lived and weighing up to 136 tons, found in oceans all round the world. The former is threatened because cultivation is progressively reducing its habitat, because it is regarded as vermin, and because it is available for target practice. The blue whale is now in the utmost danger because of senseless over-hunting in the face of considered advice from zoologists. Indeed, its southern populations, which were always the largest, have been reduced from something like 150,000 animals to fewer than 1000. In little more than 60 years its total population has been reduced by more than 99 per cent; this is surely decimation with a vengeance.

Slowly—too slowly for some species—individuals and governments are coming to realize that once an animal becomes extinct it cannot be re-created, and that the conservation of our living heritage is just as important as the husbanding of our stocks of coal, oil and iron.

Echidna

Echidna or Spiny Anteater

Tachyglossus aculeatus

Size, etc: 14-20 inches in length. Weighs up to 12 lbs. **Description:** Upper part, except the head, covered with yellowish-white, black tipped spines. Feet with five toes, armed with strong claws which it uses for digging. The animal is toothless. Lives up to 50 years. **Distribution:** Australia and New Guinea. **Food:** Insects, especially ants and termites. **Young:** Female lays single egg into a pouch in which it hatches. Young feed by sucking fur which holds milk exuded by mother's milk secreting glands. **Remarks:** Long, sticky tongue is used to catch ants and termites.

Duck-billed Platypus

Duck-billed Platypus

Ornithorhynchus anatinus

Size, etc.: Nearly 2 feet in length overall. Weighs up to 4 lbs. **Description:** Sepia-brown on back, silvery tinged with yellow or pink below. Beak-like snout, webbed feet. Tail like that of beaver. Lives up to 15 years. **Distribution:** Eastern and southern Australia, Tasmania. **Food:** Worms, crayfish, small freshwater life of which it consumes half its own weight each 24 hours. **Young:** Female usually lays two eggs about size of house-sparrow's egg and soft shelled, which take nine to ten days to incubate. At birth young are blind and naked. **Remarks:** Male has horny spur on inner of each hind legs which conducts venom secreted by a gland.

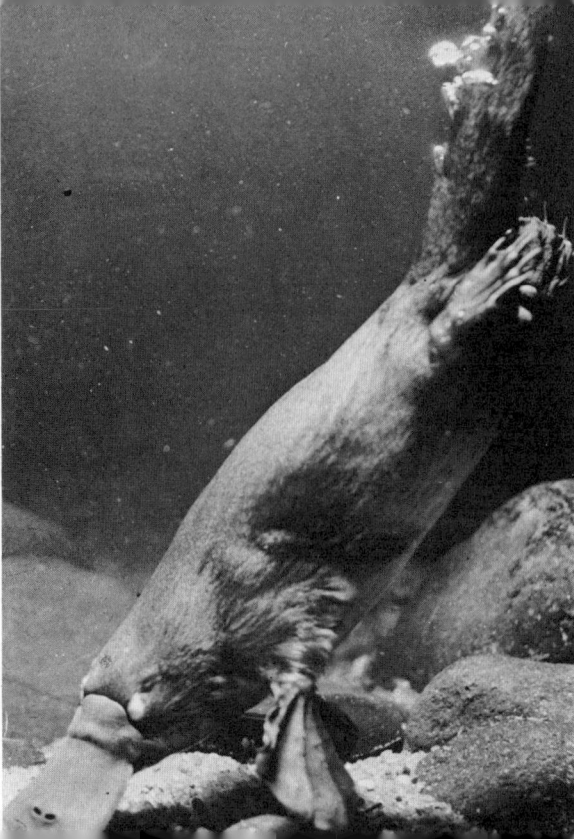

Opossum

American or Virginian Opossum

Didelphis marsupialis virginiana

Size, etc.: Length of head and body about 20 inches + 20-inch tail. **Description:** Grey-white fur. Black feet with long toes and powerful tail. Most of tail naked. Lives up to 8 years. **Distribution:** Southern Canada to Northern Argentina. **Food:** Birds, carrion, eggs and small animals. **Young:** Up to 18 at birth but only 7 survive pouch-life. Gestation period 12-13 days. **Remarks:** Sometimes feigns death when alarmed. Nocturnal.

Tasmanian Devil

Tasmanian Devil

Sarcophilus harrisii

Size, etc.: Head and body 31 inches in length + 12-inch tail. Males weigh up to 18 lbs. and females to 10 lbs. **Description:** Coarse black hairs splashed with irregular blotches of white. Stocky body. Stout claws for digging. Powerful teeth. Lives 7 to 8 years. **Distribution:** Now survives only in Tasmania. Formerly also Southern Australia. **Food:** Rats, mice frogs, birds and the like. Has been known to attack farm animals, including dogs. **Young:** 4 at birth. Pouch-life to 15 weeks of age. **Remarks:** Nocturnal.

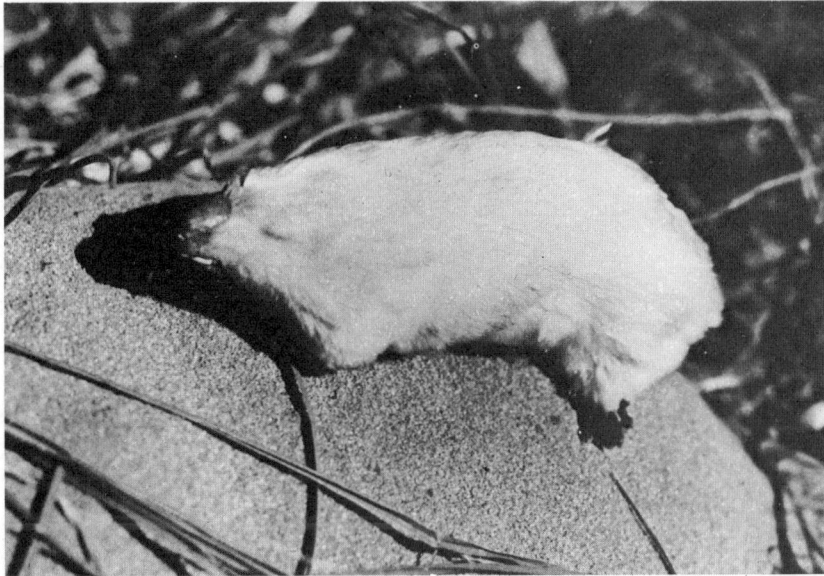

Marsupial Mole

Notoryctes typhlops

Size, etc.: 7 inches in length. Weighs up to about $3\frac{1}{2}$ ounces. **Description:** Mole-like. Eyes vestigial and hidden under skin. Silky white or gold-red fur. Five toes on each foot. Enlarged claws on front feet. Hard, horny shield on the upper part of the snout which serves as a ram in digging. **Distribution:** Northwestern and south-central Australia. **Food:** Beetles and larvae of ants, moths and butterflies. **Young:** Little is known. Female has pouch containing two teats. **Remarks:** These animals do most of their digging just below the surface and do not dig deep burrows, except possibly in breeding season.

Bandicoot

Long-nosed Bandicoot

Perameles nasuta

Size, etc.: 23 inches in length including tail. **Description:** Light brown to grey-brown fur with whitish underparts. Sharp face with large pointed ears. Long, sharp claws for digging. **Distribution:** Australia and Tasmania. **Food:** Insects, spiders, worms and even small mice. **Young:** 2-4 at birth. Gestation period at least 11 days. Carried in pouch, which opens backwards. Young leave pouch at about 55 days of age. **Remarks:** Has many enemies, including snakes, dingoes and foxes, and its natural habitat is fast being destroyed by man, sheep and cattle.

Koala or Australian Teddy Bear

Phascolarctos cinereus

Size, etc.: 24-30 inches in length. Weighs about 30 lbs. **Description:** Thick, woolly grey fur and whitish underparts. Rubbery black nose. Sharp claws. Cheek pouches for storing food. Lives about 20 years. **Distribution:** Eastern Australia. **Food:** Foliage of eucalyptus and gum tree. **Young:** 1 at birth. Carried first in pouch for six months and later on mother's back, sometimes until infant is nearly as large as its mother. Gestation period: 25-30 days. **Remarks:** Arboreal. Greatly reduced in numbers by disease and hunting, but now fully protected in the three eastern Australian states.

Wombat

Common Wombat

Vombatus ursinus

Size, etc.: 40 inches in length.
Description: Coarse fur varying from black to yellowish-buff or grizzled. Naked nose. Stout claws for digging. Lives up to 30 years.
Distribution: Southeastern Australia, Tasmania, Flinders Island. **Food:** Grasses, roots and inner bark of trees. **Young:** Bears 1 baby which is carried in mother's pouch. **Remarks:** Nocturnal, powerful burrower. Quick in its movements, it can also run quickly, at least for short distances.

Wallaby

Brush-tailed Rock Wallaby

Petrogale penicillata

Size, etc.: 40-60 inches in length including tail. **Description:** Coats of individuals vary considerably in colour, from brown to a bright orange. Hind feet padded to prevent animal slipping on rock faces. **Distribution:** Australia. **Food:** Vegetation, including grasses, leaves, roots, berries, etc. **Young:** 1 at birth. Carried in mother's pouch. **Remarks:** Nocturnal, spending day in rock crevices. Travels awkwardly in open country.

Kangaroo

Great Grey Kangaroo

Macropus canguru

Size, etc.: 6-7 feet in length + 4-foot tail. Weighs up to about 140 lbs. **Description:** Grey with white underparts. Can leap 25 feet (one jump recorded measured 38 feet) and can travel at 25 m.p.h. when necessary. Lives up to 15 years. **Distribution:** Australia and Tasmania. **Food:** Vegetation, grasses, roots, berries, etc. **Young:** 1 at birth. Carried in mother's pouch. Gestation period 30-40 days. **Remarks:** Feeds throughout the night and rests during the heat of the day.

Wallaroo

Wallaroo or Euro

Macropus robustus

Size, etc.: About 48 inches in length including tail. **Description:** Dark grey with whitish chest and hairless muzzle. Short broad feet. **Distribution:** Australia, mainly in coastal mountains and rocky inland ranges. **Food:** Vegetation, including grasses, leaves and roots. **Young:** 1 at birth. Carried in mother's pouch. **Remarks:** Can go without water for two to three months. Sometimes uses caves as shelters.

Tree Kangaroo

Tree Kangaroo

Dendrolagus spp *(5 species)*

Size, etc.: 30-40 inches in length including tail. **Description:** Coat varies from near black to fawn. Long hind legs and slender tail with brush at tip. **Distribution:** New Guinea and northeastern Queensland. **Food:** Leaves, ferns, fruits, etc. **Young:** 1 at birth. Carried in mother's pouch. **Remarks:** Climbs trees to feed and sleep. Can leap 50 feet or more to ground.

Common Tenrec

Centetes ecaudatus

Size, etc.: 12-16 inches in length. **Description:** Soft yellowish-brown fur. Collar of longer hairs round neck which bristle when attacked. Long snout. Tailless. **Distribution:** Madagascar and the Comoro Islands. **Food:** Insects, worms and roots. **Young:** 15-21 at birth. **Remarks:** Sleeps in burrows by day and forages in the evening and at night. Shelters in hollow logs or under rocks.

Hedgehog

Hedgehog

Erinaceus europaeus

Size, etc.: 7-9 inches in length.
Description: Chocolate-brown with tips of spines yellowish-white. Face, legs and belly covered with coarse hair. Short legs with strong claws for digging. Lives up to 6 years.
Distribution: Europe and Asia as far as Korea. Introduced to New Zealand. **Food:** Insects, roots, worms, small mammals, frogs and lizards. **Young:** 3-7 at birth. Two litters per year. Gestation period 34-48 days. **Remarks:** Rolls itself into a ball at least sign of danger. Is a good climber and swimmer. Hibernates during winter.

Common Shrew

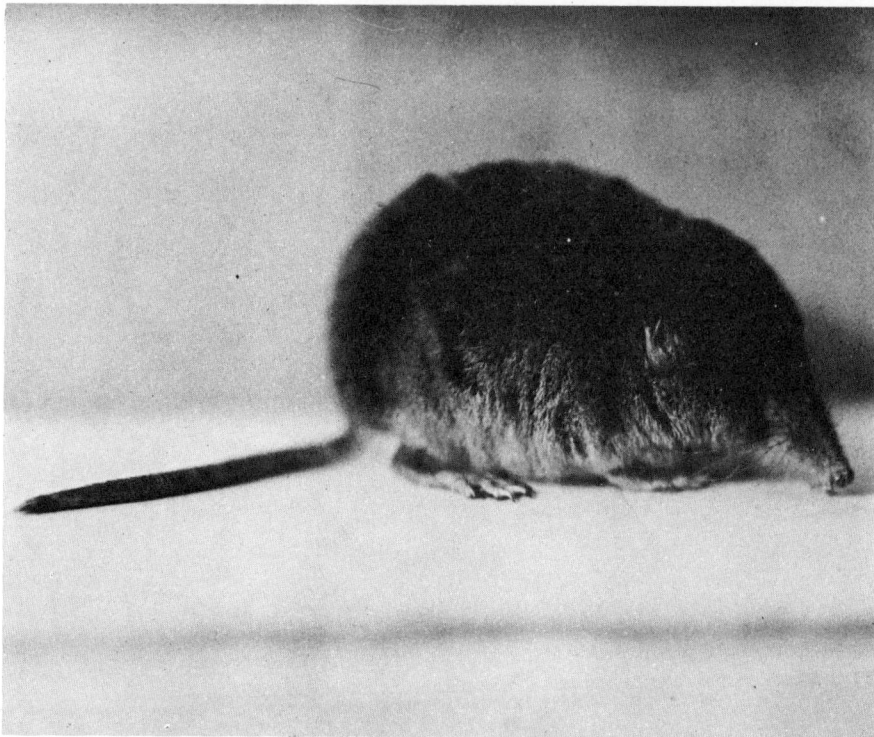

Common Shrew

Sorex araneus

Size, etc.: 2-3 inches + 1-2-inch tail. **Description:** Dark brown coat of short, silky fur. Short limbs with 5 toes. Pointed, whiskered snout. Needle-sharp teeth. Lives up to 15 months. **Distribution:** Western Europe and Asia. **Food:** Wide range of animal and vegetable matter. Possibly eats about its own weight in food every 24 hours. **Young:** 4-10 at birth. 3 to 4 broods per year. Gestation period 18-28 days. **Remarks:** Saliva contains venom, which incapacitates its prey.

Water Shrew

Water Shrew

Neomys fodiens

Size, etc.: About 3 inches + 3-inch tail. **Description:** Dark brownish coat with silvery white underparts. Tail has keel of stiff hairs on underside. Lives perhaps 14-19 months. **Distribution:** Rivers of Europe and northern Asia. **Food:** Water insects, frogs and fresh-water crustaceans. **Young:** 4-8 at birth. Gestation period 24 days. Born blind and naked. **Remarks:** Excellent swimmer but does not have webbed feet. Produces a poisonous secretion from salivary glands, which is used to weaken its prey.

Common Mole

Talpa europaea

Size, etc.: 5-6 inches + 1-inch tail. Weighs up to about 4 ozs. **Description:** Black velvety fur, sometimes with tinge of grey or blue. Short legs and powerful feet, particularly forefeet, for digging. Lives about 3 years. **Distribution:** Europe and Asia to Japan. **Food:** Mainly earthworms, also insects and small mammals. Must eat at frequent intervals. Dies if deprived of food for 10 hours. **Young:** 2-6 at birth. Young born naked and blind. Sometimes a second litter. Gestation period 42 days. **Remarks:** A surprisingly good swimmer. Eyes minute.

Fruit Bat

Fruit Bat or Flying Fox

Pteropus spp (35 species)

Size, etc.: 12 inches with wing span of 45-50 inches. Weighs up to about 2 lbs. **Description:** Foxlike head. Large eyes. Black wings. Brownish-red fur, sometimes a dirty yellow colour. Lives to at least 17 years of age. **Distribution:** India, Madagascar, Australia and Pacific Islands. **Food:** Fruits, wild and cultivated. **Young:** 1 at birth. Gestation period 180 days. **Remarks:** Navigates by means of sound radar. Has a strong smell. Roosts in large colonies. Feeds at dusk.

Vampire or Blood-Sucking Bat

Desmodus rotundus

Size, etc.: About 3 inches in length. Weighs almost 2 ozs. **Description:** Drab grey or brownish short hairs. Low pointed ears. **Distribution:** Central and South America. **Food:** Blood of various animals, including man. **Young:** 1 at birth. Gestation period 90-120 days. **Remarks:** Makes incision with razor-sharp front teeth and laps up blood. Can move quite swiftly on the ground on its legs and forearms. Is a carrier of rabies.

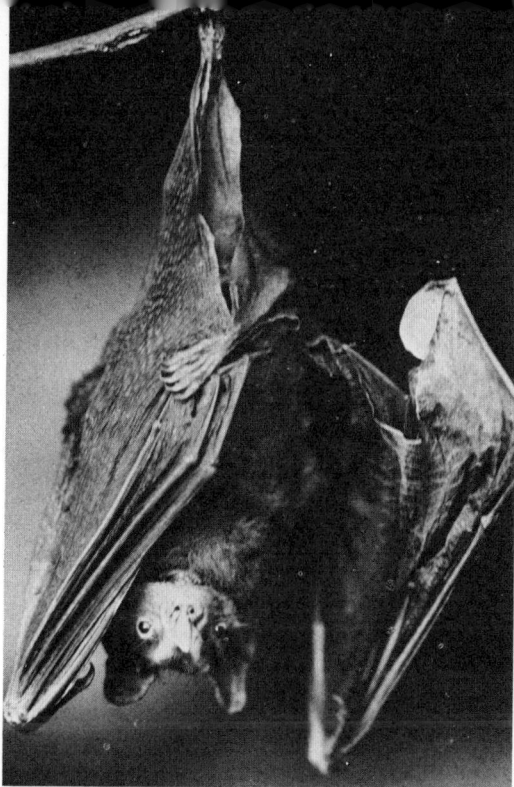

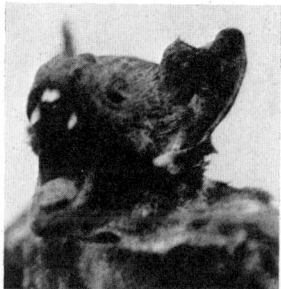

Long-Eared Bat

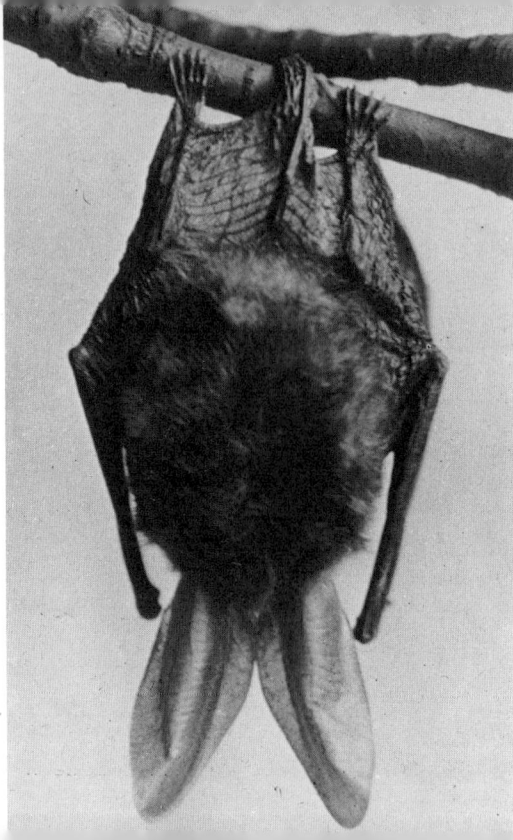

Long-eared Bat

Plecotus auritus

Size, etc.: 3 inches + 2-inch tail. Wing span about 10 inches. Large ears up to $1\frac{1}{2}$ inches long. **Description:** Soft brown fur with lighter underparts. Has a delicate butterfly style of flight and can hover rather like a humming bird. Lives up to $12\frac{1}{2}$ years. **Distribution:** Europe, North Africa and Asia as far as Japan. **Food:** Insects, especially small moths. **Young:** 1 at birth. **Remarks:** Hibernates October-April. Does not begin flying until after dark.

Ring-Tailed Lemur

Ring-tailed Lemur

Lemur catta

Size, etc.: 45 inches including tail. **Description:** Grey fur. Black and white ringed tail. Fox-like face with horizontal incisor teeth. Lives for more than 25 years. **Distribution:** Madagascar and the Comoro Islands. **Food:** Fruit and insects. **Young:** 1 at birth. Gestation period about 40 days. **Remarks:** A night feeder. Terrestrial, lives among rocks in thinly-wooded country.

Slow Loris

Slow Loris

Nycticebus coucang

Size, etc.: 12-14 inches in length. Weighs up to about 3 lbs. **Description:** Thick woolly silver-grey or fawn fur. Dark stripes on head. Tailless. Lives up to 10 years. **Distribution:** Philippines, East Indies, and tropical regions of Southern Asia. **Food:** Insects, leaves, fruit and small animals. **Young:** 1 at birth. Gestation period about 90 days. **Remarks:** Notable for its "slow-motion" movements. Rarely descends from trees.

Bushbaby or Senegal Galago

Galago senegalensis

Size, etc.: 10-12 inches in length.
Description: Thick woolly brown
fur and reddish limbs. Large trans-
lucent eyes. Slender toes and fingers
equipped with fleshy pads. **Distribution:** Equatorial and southern
Africa. **Food:** Insects, leaves, fruits
and eggs. **Young:** 1 or 2 young at
birth. Gestation period 120 days.
Remarks: Pads on fingers and toes
give the Bushbaby complete freedom
in trees. Able to jump 20 or more
feet horizontally.

Tarsier

Tarsier

Tarsius spectrum

Size, etc.: 5-6 inches + 10-inch tail. Weighs about 5 ozs. **Description:** Thick woolly brownish-grey fur. Round face with large owl-like eyes. Fingers and toes end with fleshy pads. **Distribution:** Sumatra, Philippines and Celebes. **Food:** Insects, lizards, small birds and fruits. **Young:** Thought to bear a single baby yearly. **Remarks:** A night feeder. Makes flying leaps from tree to tree.

Douroucouli

Douroucouli or Night Monkey

Aotus trivirgatus

Size, etc.: 14-inch body + tail fractionally longer. **Description:** Rusty brown or reddish-brown fur and furry tail. Large saucer-like eyes. May live to 20 years of age. **Distribution:** Central America, Brazil, Peru and Ecuador. **Food:** Fruit and insects. **Young:** 1 at birth. **Remarks:** Has a caterwauling cry which has led to it being called the "devil-monkey". Night feeding. Lives in troops.

Squirrel Monkey

Squirrel Monkey

Saimiri sciureus

Size, etc.: About 12 inches + 12-15-inch tail. Weighs up to about 2 lbs. **Description:** Back, hands and feet a vivid orange-yellow, with shoulders, legs and underparts yellowish-grey. Sometimes has a crown of black or grey on the top of its head. White around eyes, ears, on throat and sides of neck. May live to 20 years of age. **Distribution:** Costa Rica to Bolivia and Peru. **Food:** Insects, fruit, lizards, small birds and eggs. **Young:** 1 at birth. **Remarks:** Feeds during the daytime.

Brown Capuchin

Cebus apella

Size, etc.: 14-20 inches + 20-inch prehensile tail. **Description:** Black or brown hair on body but often yellowish or golden buff on chest, shoulders and sides of face. Fairly long arms and legs. Hairs on head form "peak", rather like a monk's hood, hence its name. Long prehensile tail. **Distribution:** Honduras to Paraguay. **Food:** Insects, fruits, nestling birds and eggs. **Young:** 1 baby at birth. Gestation period 180 days. **Remarks:** The Capuchin is the most common monkey of South America.

Humboldt's Woolly Monkey

Humboldt's Woolly Monkey

Lagothrix humboldtii

Size, etc.: 18-24 inches + 24-inch prehensile tail. **Description:** Dense dark grey woolly fur. Naked face which has a sad appearance. Head and face are black. **Distribution:** Amazonia and Orinoco. **Food:** Mainly fruit-eater but sometimes leaves and flowers, and raw flesh. **Young:** 1 at birth. Gestation period about 5 months. **Remarks:** Feeds during the daytime. Lives in troops.

Pinchè Marmoset

Pinchè Marmoset or Cotton-headed Tamarin

Saguinus oedipus

Size, etc.: 10-12 inches + 15-inch tail. **Description:** Brown coat with long head-cape. Long tail. 32 teeth. **Distribution:** South America to Southern Brazil and Peru. **Food:** Fruit and insects. **Young:** 1 at birth. Gestation period about 5 months. **Remarks:** Has bird-like voice of flute-sounding notes and trills.

Baboon

Chacma Baboon

Papio ursinus

Size, etc.: 24 inches at shoulder. Length 48 inches including 18-inch tail. **Description:** Dog-like muzzle. Grey-brown coat. Purple face. Lives perhaps up to 10 years in wild, but has lived to 45 years in captivity. **Distribution:** Eastern and Southern Africa. **Food:** Fruit, grain, insects. **Young:** 1-2 at birth. Gestation period 173-193 days. **Remarks:** Feeds during the daytime. Lives in large troops. Steals cultivated crops.

De Brazza's Monkey

De Brazza's Monkey

Cercopithecus neglectus

Size, etc.: 21 inches + 30-inch tail. **Description:** Long tail. Red-yellow band on forehead. Black band above. White chin and cheeks. **Distribution:** Rain forests of Equatorial Africa. **Food:** Eggs, fruits, insects. **Young:** Usually one young, occasionally twins. Gestation period about 7 months. **Remarks:** Lives in troops. Steals cultivated crops.

Langur

Capped Langur or Helmeted Langur

Presbytis entellus

Size, etc.: 25-30 inches + 40-inch tail. Weighs up to 40 lbs. **Description:** Red, or red and black. Long hands and fingers. **Distribution:** Burma, Malaya, Thailand. **Food:** Fruits and leaves. **Young:** 1 at birth. Gestation period about 196 days. **Remarks:** Arboreal but can move across ground very quickly for a primate.

Proboscis Monkey

Proboscis Monkey

Nasalis larvatus

Size, etc.: 30-inches + 30-inch tail. Male weighs up to 44 lbs., female about half this weight. **Description:** Orange-brownish coat, darker on top of head and back, and lighter almost to white below. Male has long "snozzle"-like nose, hence its name. Female has shorter turned up or "retrousé" nose. **Distribution:** Borneo. **Food:** Fruit and leaf eater. **Young:** 1 at birth. **Remarks:** Usually found in low mangrove swamps, delta regions, and river banks.

Gibbon

Hoolock Gibbon

Hylobates hoolock

Size, etc.: Head and body 18-25 inches. Weighs up to 16 lbs. **Description:** Male has black coat, female light brown. Delicately built. Its long arms enable it to make 35-40 foot swings from tree to tree. Lives up to 23 years. **Distribution:** Burma and Assam. **Food:** Fruit, leaves and buds, occasionally insects, young birds and eggs. **Young:** 1 at birth. Gestation period 210 days. **Remarks:** Walks erect when on ground. Travels in troops. Exceeds all other animals in agility.

Gorilla

Gorilla gorilla

Size, etc.: Height 5½ feet. Arm-spread 8 feet. Weighs up to 550 lbs (adult male). **Description:** Black to steel-grey coat. Male has a crested topknot. Lives up to 50 years. **Distribution:** Equatorial Africa. **Food:** Fruits and vegetables. **Young:** 1 at birth. Gestation period 270 days. **Remarks:** For all its power, basically a peaceful creature and never kills. Usually walks on all fours, leaning on its knuckles. Travels in family groups. Numbers greatly reduced, particularly of Mountain gorilla subsp, which is regarded as dangerously threatened.

Giant Anteater

Giant Anteater

Myrmecophaga tridactyla

Size, etc.: 2 feet in height, 8 feet in length including tail. Weighs up to 46 lbs. **Description:** Grizzled coarse fur marked with distinct broad black strips on collar and back, with white extending from throat over shoulders and ending on back. Long nozzle-shaped head. Toothless. A 12-inch tongue. Powerful claws for tearing down termite hills and for self-defence. Has lived 14 years in captivity. **Distribution:** Tropical forests of Central and South America. **Food:** Termites and ants. **Young:** 1 at birth. Gestation period about 190 days. **Remarks:** Tongue, which is coated with sticky saliva to which termites, ants and other insects adhere, can be extended up to 24 inches. Diurnal in the wild, but nocturnal in densely populated areas.

Two-Toed Sloth

Two-toed Sloth

Choloepus didactylus

Size, etc.: 27 inches in length. Weighs about 18 lbs. **Description:** Olive-brown coat which sometimes takes on a greenish colour, from the alga which grows in it, and has the appearance of matted grass. No tail. Has 2 claws on forefeet and 3 on hind. Has lived 11 years in captivity. **Distribution:** Costa Rica to Ecuador. **Food:** Fruit and leaf buds. **Young:** 1 at birth. Gestation period about 263 days. **Remarks:** Almost helpless on the ground, the sloth spends its time upside-down in trees. Swims voluntarily, using the breast-stroke with the body right way up.

Armadillo

Nine-banded Armadillo

Dasypus novemcinctus

Size, etc.: 18-20 inches + 12-inch tail. **Description:** Greyish-brown. Nine movable bands about its middle, separating front and rear casing and so allowing animal to curl into ball. Pointed face and ears. Tail made up of interlocking horn rings. Powerful front and hind claws. **Distribution:** From southern United States to Argentina. **Food:** Ants, beetles and other small insects. **Young:** 4-10 at birth. Gestation period about 260 days. **Remarks:** A good swimmer. Feeds mainly at night. Gregarious, several sharing same burrow.

Pangolin

Giant African Pangolin

Manis gigantea

Size, etc.: 60 inches including tail. **Description:** Grey-brown. Large scales. Underparts bare of scales. Tail almost as long as body. Protractile tongue. Small eyes protected by heavy thick eyelids. **Distribution:** Forest and open country of East, Central and southern Africa. **Food:** Ants and termites. **Young:** 1 at birth. **Remarks:** Terrestrial and nocturnal. Rolls into ball with scales on outside, for defence.

Mouse-Hare

Pika or Mouse-Hare

Ochotona princeps

Size, etc.: About 6 inches in length, and weighs 3-4 ozs. **Description:** Reddish-brown soft fur. Looks rather like a miniature rabbit, but has short rounded ears. **Distribution:** Mountainous regions of eastern Europe, Asia (but not south of Himalayas or SE Asia) and western North America. **Food:** Green vegetables and grasses. In late summer gathers grasses, sedges and twigs and lays them out in the sun to dry for hay. This is then eaten during the winter. **Young:** 3-5 at birth. Gestation period 30 days. Life span seems to be one to three years. **Remarks:** Lives in colonies. Feeds during the day. Twelve species in the old world and two in the New World.

Brown Hare

Brown Hare

Lepus europaeus

Size, etc.: 24 inches + 3-4-inch tail. Weighs about 8 lbs. **Description:** Greyish-brown fur. Very long ears. Strongly developed hind-legs. Lives up to 12 years. **Distribution:** Europe and Asia. **Food:** Vegetables and grasses. **Young:** 2-4 at birth. **Remarks:** A healthy adult can run at a speed of 45 m.p.h. Lives in a hollow in grass on the surface of the ground and *not* in burrows.

Jack Rabbit

Jack Rabbit

Lepus californicus

Size, etc.: About 24 inches + 3-4-inch tail. **Description:** Greyish-brown fur with patches of white on head and ears. Black tail. Long ears and large hind feet. **Distribution:** Grass plains of western United States. **Food:** Grasses and vegetables. **Young:** 1-6 at birth. Gestation period 41-47 days. **Remarks:** This species is a true hare and not a rabbit. Does not burrow. Spends inactivity hidden in vegetation. Mainly nocturnal.

Common Rabbit

Oryctolagus cuniculus

Size, etc.: 16-18 inches in length. Weighs up to 5 lbs. **Description:** Greyish-brown fur. Longish ears. Short tail with white underside. **Distribution:** Europe and North Africa. Introduced to Australia where it breeds profusely. Also introduced to New Zealand, South America and parts of North America. **Food:** Vegetables, grasses and bark. **Young:** 3-9 at birth. Gestation period 28-33 days. **Remarks:** Lives in communal burrows or warrens.

Marmot

Alpine Marmot

Marmota marmota

Size, etc.: About 24 inches + 4½-inch tail. Height 7 inches at shoulder. Lives 13-15 years. **Description:** Reddish-yellow coat with greyish head. Thumbless. **Distribution:** Mountains of south-east Europe and Asia. **Food:** Grasses and vegetables. **Young:** 4-6 at birth. Gestation period 35 to 42 days. **Remarks:** In summer lives in pairs in burrows just below surface of ground. In winter hibernates in large colonies for six months of the year. There are about 16 species of marmots or woodchucks.

Flying Squirrel

American Flying Squirrel

Glaucomys volans

Size, etc.: 12 inches + 4-inch tail. **Description:** Brown coat on back and sides with whitish-cream belly. Rather large eyes. Lives up to 10 years. **Distribution:** Eastern United States, Mexico and Guatemala. **Food:** Insects, fruits, leaves and small animals. **Young:** 2-6 at birth. Gestation period 40 days. **Remarks:** Feeds at night. Glides by means of skin folds stretched between wrists and ankles along each side of the body.

Pocket Gopher

Eastern American Pocket Gopher

Geomys bursarius

Size, etc.: 10-15 inches including short tail. **Description:** Coarse coat, usually a dull brown. Short sturdy legs and thickset body. Forefeet armed with powerful claws for digging. Pair of large chisel-like teeth in upper and lower jaws. Large cheek pouches or "pockets" for storing food at side of face. **Distribution:** Plains of North and Central America. **Food:** Roots, tubers and grass stems. **Young:** 1-3 at birth. Gestation period 18-19 days. **Remarks:** Spends almost all its time underground.

Beaver

Beaver

Castor canadensis

Size, etc.: 3-4 feet including 12-inch tail. **Description:** Coarse greyish-brown coat. Flat naked tail. Small forefeet and large, webbed hind feet. Powerful incisor teeth which it uses for felling trees. Lives up to 19 years. **Distribution:** Alaska, Canada and U.S.A. as far south as Rio Grande. **Food:** Tree bark, leaves, roots and berries. **Young:** 1-8 at birth. Gestation period 60-128 days. **Remarks:** Can remain under water for 15 minutes or more. Beavers build lodges up to 40 feet in diameter and 4-5 feet high. They also construct dams (4-5 feet high and several hundred feet long — maximum 1800 feet) to form ponds.

Hamster

Common or Black-Bellied Hamster

Cricetus cricetus

Size, etc.: 10-12 inches. Weighs up to almost 2 lbs. **Description:** Thick fur, light brown on back, white and red markings on sides with black underparts. Short limbs and delicate claws. Lives up to 2½ years. **Distribution:** Europe and Western Asia into Siberia. **Food:** Roots, grains, etc., but is known to kill and eat small animals. **Young:** 5-18 at birth. Gestation period 16-20 days. **Remarks:** Lives in burrows. Hibernates during the winter but, when it wakes, eats from its cache of food. May store up to 180 lbs of seeds, peas and potatoes in its burrow.

Water Vole

Arvicola terrestris

Size, etc.: About 10 inches including 4-inch tail. **Description:** Brownish-grey fluffy thick fur and shortish hairy tail. Makes nest in ball of grass just above water level or hole in bank. Good swimmer and diver; does not have webbed feet. **Distribution:** Europe, parts of Russia and Siberia, Asia Minor, Northern Syria, Israel and Iran. **Food:** Mainly water vegetation but on occasion small fish and fresh-water crustaceans. **Young:** 2-7 at birth. Gestation period said to be 42 days. 3-4 litters per year. **Remarks:** Does not hibernate, but burrows under the snow. Generally good swimmer but some individuals more aquatic than others.

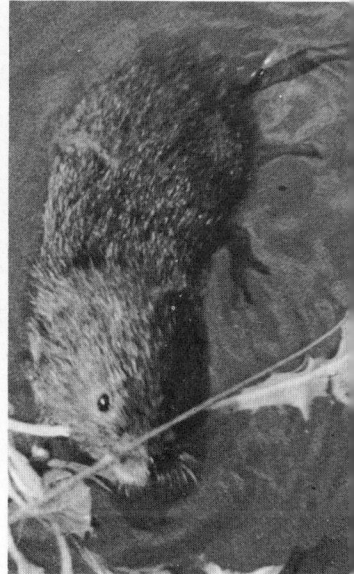

Long-Tailed Field-Mouse

Long-tailed Field-mouse or Wood Mouse

Apodemus sylvaticus

Size, etc.: 3-4 inches + 3-4-inch tail. **Description:** Drab-coloured soft fur. Sharp black eyes. **Distribution:** Most of the Northern Hemisphere. **Food:** Leaves, berries, seeds and grasses, and insects. **Young:** 5-9 at birth. 3-4 litters per year. Gestation period 21-29 days. **Remarks:** Eats its own weight in food every 24 hours.

Brown Rat

Brown Rat or Norway Rat

Rattus norvegicus

Size, etc.: 9-10 inches + 8-9-inch tail. **Description:** Brownish-grey coat with dirty-white underparts. Flesh-coloured feet, tail and ears. Lives up to 5 years. **Distribution:** World-wide—excepting Arctic regions. **Food:** Practically anything it can cut with its teeth. **Young:** 4-8 at birth. 6 litters per year. Gestation period 20-26 days. **Remarks:** Often associated with populated regions. Is an expert swimmer. Genus contains at least 570 named species.

Dormouse

Dormouse

Muscardinus avellanarius

Size, etc.: 5-6 inches including tail.
Description: Light brown fur above with whitish-yellow underparts. Tail well haired. Lives up to 3 years.
Distribution: Hedgerows and woods of Europe and Western Asia.
Food: Nuts, insects and eggs.
Young: 2-9 at birth. 3-4 litters per year. Gestation period 21-28 days. Born blind and naked. **Remarks:** Nocturnal. Hibernates autumn to spring. Arboreal but limits climbing to the lower branches of trees or shrubs.

Jerboa

Size, etc.: About 7 inches + 8-inch tail. **Description:** Various shades of buff, depending on locality, with black or pale russet and white underparts. Highly developed hind legs and small short front legs. Long tail carries a distinctive black and white brush of hairs. Lives 5½-6½ years. **Distribution:** Deserts of northern Africa and southwestern Asia. **Food:** Grass seeds, vegetable matter and insects. **Young:** Little is known of breeding habits. 3-4 young in a litter. Gestation period of about 40 days. **Remarks:** Can leap as much as 6 feet but normally takes leaps of about 2 feet. Lives in burrows and may emerge on moonlight nights to play and hold "cocktail parties".

65

Porcupine

Common or Crested Porcupine

Hystrix cristata

Size, etc.: Up to 33 inches in length. Weighs up to 45 lbs. **Description:** Black, brown and white quills on back and tail. Crest of bristles on head and neck. Lives for 8-12 years, but may reach 20 years. **Distribution:** Hills of Italy and tropical Africa. **Food:** Fruit and roots. **Young:** 2-3 at birth. Gestation period 63-112 days. **Remarks:** Is the largest African rodent. Erects its quills for defence. Almost completely nocturnal, spending the day in a burrow, in natural cavity or crevice.

Guinea Pig

Guinea Pig

Cavia porcellus

Size, etc.: Up to 12 inches in length. **Description:** Reddish- to greyish-brown fur in wild state, but varies in domestic animals. Tailless. Rather fat body and short limbs. Lives up to 8 years. **Distribution:** South America. Introduced to Europe by traders. **Food:** Vegetable matter. **Young:** 4-12 at birth. Gestation period 63-71 days. **Remarks:** Lives in burrows of its own construction, or occupies abandoned ones of other animals. Largely nocturnal.

Capybara

Capybara or Water Cavy

Hydrochoerus hydrochaeris

Size, etc.: 4 feet in length, height 20-22 inches at the shoulder. Weighs up to 125 lbs. **Description:** Stiff yellowish-brown fur. Broad, blunt head with rather small eyes. Short legs and only the trace of a tail. Lives up to 10 years. **Distribution:** Woods with dense vegetation around lakes and rivers of Central and South America. **Food:** Grasses and water plants. **Young:** 2-8 at birth. Gestation period 119-126 days. **Remarks:** Is the largest living rodent. Active morning and evening.

Golden Agouti or Aguti

Dasyprocta aguti

Size, etc.: 16-20 inches in length. Weighs 2½-8 lbs. **Description:** Sleek golden-brown or reddish coat. Slender legs each with 3 toes bearing hoof-like claws. Lives 13-20 years. **Distribution:** Forests, thick brush, dry hillsides and cultivated areas of tropical America and West Indies. **Food:** Leaves, roots and fallen fruits. **Young:** 2-4 at birth, born in burrow after gestation period of 64 days. 2 litters per year. **Remarks:** Is a swift runner and great leaper (up to 20 feet has been recorded). There are about 24 closely related species in the genus.

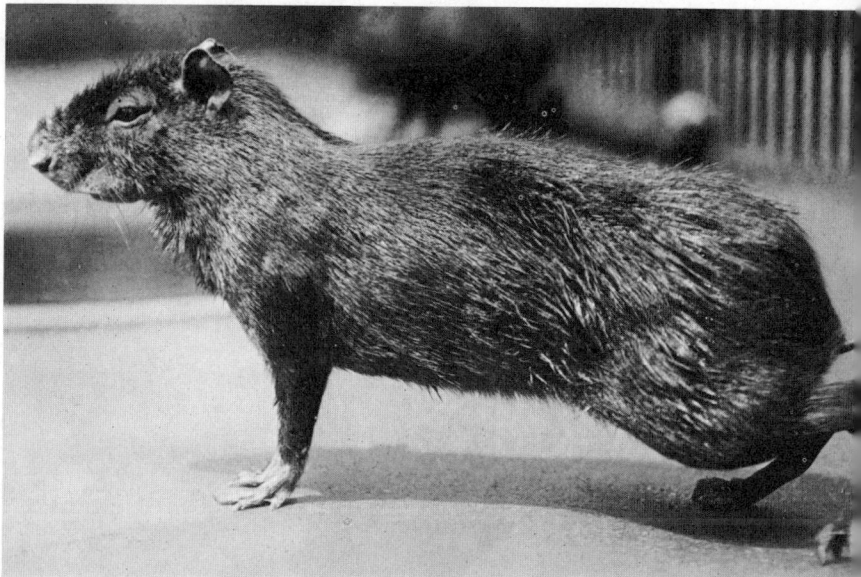

Chinchilla

Chinchilla

Chinchilla laniger

Size, etc.: Up to 20 inches including its bushy tail. **Description:** Pale blue-grey or silver-grey coat which is highly regarded by the fur trade. Has soft pads on the ends of its toes to protect them. Lives up to 10 years but 20 years in captivity. **Distribution:** Andes of Chile and Bolivia. **Food:** Any available vegetation. **Young:** 1-6 at birth. 3 litters a year. Gestation period 105-114 days. **Remarks:** Formerly almost exterminated by trapping, but now increasing as a result of protection.

Coypu

Coypu

Myocastor coypus

Size, etc.: 26 inches in length + 15 inch tail. Weighs up to 20 lbs. **Description:** Buffish-brown thick heavy fur. Dense yellow velvety underfur—the "nutria" used by the fur trade. Lives up to 6 years. **Distribution:** Central and southern South America. Introduced into United States and Europe, especially eastern England. **Food:** Green vegetation, frequently eat molluscs. **Young:** 3-9 at birth. Gestation period 120-150 days. Weaned in 7-8 weeks. Two or three litters a year. **Remarks:** Lives a great part of its life in water and builds burrows in river banks. Female's teats are high up on its side to allow young to feed while on female's back, or when she is lying on her belly.

Cane Rat

Cane Rat

Thryonomys swinderianus

Size, etc.: 24 inches + 6-7-inch tapering tail. Weighs up to 18 lbs. **Description:** Dense grey-brown spiny hair with lighter underside. Lives to about three years. **Distribution:** Africa, south of the Sahara. **Food:** Water grass, reed roots, bark and other green vegetation. **Young:** 2-4 young at birth. **Remarks:** Is a strong swimmer. Usually found near water.

Sperm Whale

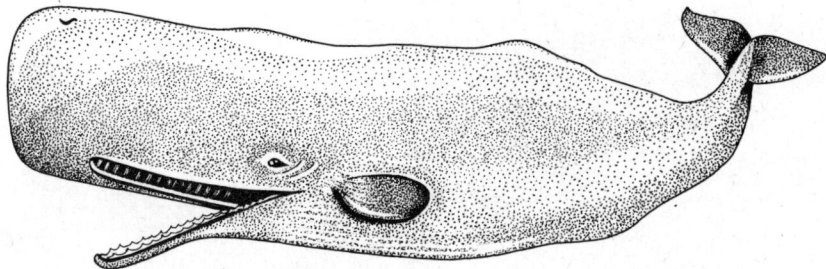

Sperm Whale

Physeter catodon

Size, etc.: Male 60 feet in length. Female about half as large (head occupies more than $\frac{1}{3}$ total length). Weighs 35 to 50 tons. Functional teeth only in lower jaw. **Description:** Black back and sides. Belly somewhat lighter. **Distribution:** Oceans throughout the world. **Food:** Cuttlefish and squid. **Young:** 1 calf at birth. Gestation period 356-480 days. **Remarks:** The largest toothed whale. Yields spermaceti, sperm oil and ambergris. Gregarious and polygamous. Migratory. Feed at great depths, often at 1,000 ft. and sometimes to over 3,200 ft. Can remain submerged for at least 45 minutes.

Dolphin

Common Dolphin

Delphinus delphis

Size, etc.: 8 feet in length. Weighs up to 150 lbs. **Description:** Black or dark brown. White belly. Head is beaked. Lives 25-30 years. **Distribution:** Temperate and warm seas throughout the world. **Food:** Fish. **Young:** 1 calf at birth. Gestation period about 276 days. **Remarks:** Capable of swimming at more than 25 mph, and usually travels in groups of 20 to several hundred individuals.

Killer Whale

Killer Whale or Grampus

Orcinus orca

Size, etc.: 30 feet in length. **Description:** Black with white belly. White patch above eye. Triangular dorsal fin (6 ft high in males). **Distribution:** World-wide, but especially polar regions. **Food:** Porpoises, sea-birds, seals, walrus, whalebone whales, fish. **Young:** 1 at birth. Gestation period about one year. **Remarks:** Is reputedly the most-feared sea creature, but this may be undeserved as they appear to be amenable in captivity.

Common Porpoise

Common Porpoise

Phocaena phocaena

Size, etc.: 4-6 feet in length. Weighs up to 120 lbs. **Description:** Black back. White belly. Black flippers. **Distribution:** Coastal regions of North Atlantic and north-east Pacific. **Food:** Fish, crustaceans and cuttle-fish. **Young:** 1 calf at birth. Gestation period about 11 months. **Remarks:** Conical head is not beaked. Usually frequents coasts and mouths of large rivers. Less playful than most porpoises and dolphins.

Sei Whale

Sei Whale

Balaenoptera borealis

Size, etc.: About 60 ft in length and probably weighs a maximum of about 55 tons. **Description:** Dark blue-grey overall except that somewhat paler on underside of front half of body. Whalebone has grey-white, soft fine texture to inner frayed margins. **Distribution:** Widely distributed in all oceans, as far as edge of ice in polar latitudes, to which it migrates in summer. **Food:** Planktonic shrimp-like crustaceans (krill). **Young:** A single calf is born at a time following a gestation period of 12 months. Females usually bear young in alternate years, and suckling lasts about six months. **Remarks:** Although small, Sei whales are being extensively hunted following the decline in the numbers of the commercially more valuable blue and fin whales.

Right Whale

Greenland Right Whale

Balaena mysticetus

Size, etc.: 60 feet in length. (Head occupies nearly $\frac{1}{3}$ of total length). **Description:** Black, white patch on chin. No dorsal fin. Baleen black. **Distribution:** Arctic regions of Atlantic and Pacific. **Food:** Shrimp-like crustaceans (krill). **Young:** 1 calf at birth (15 feet long). Gestation period perhaps 276 days, but may be up to about one year. **Remarks:** Baleen up to 14 ft in length. Formerly hunted but has been protected for many years.

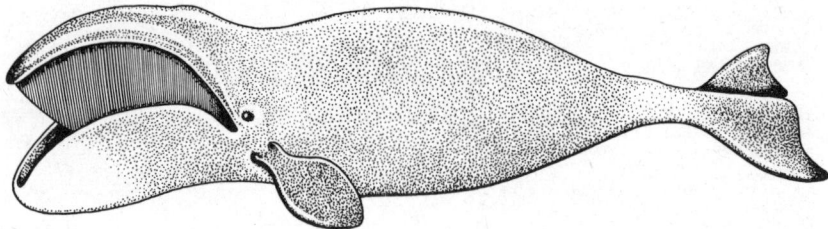

Asiatic Jackal

Golden or Asiatic Jackal

Canis aureus

Size, etc.: 30 inches in length. Height 16 inches at shoulder. 8-10-inch tail. Weighs about 20 lbs. **Description:** Grey coat of long hairs and bushy tail. Lives up to 16 years. **Distribution:** Central Africa, Middle and Far East. **Food:** Carrion and small mammals. **Young:** 3-5 at birth. Gestation period 60-63 days. **Remarks:** Is no mean runner and credited with speed of 35 m.p.h. Has long eerie wail. Is a night feeder.

Common Wolf

Common Wolf

Canis lupus

Size, etc.: About 50 inches, 30 inches at the shoulder. 15-16 inch tail. Weighs 115 lbs on average. **Description:** Coats vary in colour from almost black to white, some brindled brown or yellow. Lives up to 17 years. **Distribution:** Northern Hemisphere. **Food:** Mammals up to size of horses. **Young:** 3-12 at birth. Gestation period 60-63 days. **Remarks:** Is a savage hunter. On arctic coast of Alaska and Western Canada the coat is white throughout the year.

Arctic Fox

Arctic or White Fox

Alopex lagopus

Size, etc.: 20 inches + 10-inch tail. **Description:** Dense woolly grey-brown coat in summer and snow-white in winter. Also blue form which is dark bluish-grey in summer and paler in winter. Pointed face with short furry ears. Lives up to 14 years. **Distribution:** Arctic regions of New and Old World. **Food:** Stranded fish, lemmings, birds and eggs. **Young:** 4-11 cubs at birth. Gestation period 51-57 days. **Remarks:** Numbers fluctuate parallel with those of lemmings on which it preys.

Red Fox

Red Fox or Common Fox

Vulpes vulpes

Size, etc.: 30 inches in length + 17-inch tail. 15-16 inches at the shoulder. **Description:** Golden to brownish-red coat with white underparts and black ears and legs. Pointed face and erect ears. **Distribution:** Northern Hemisphere. **Food:** Small mammals and birds, domestic animals especially poultry, insects, grasses, fruits, etc. **Young:** 3-7 at birth. Gestation period 49-56 days. **Remarks:** A solitary creature, except in mating season. Hunts mainly at night.

Polar Bear

Polar Bear

Thalarctos maritimus

Size, etc.: 7-9 feet in length. Height 5 feet at the shoulder. Weighs an average of 900 lbs. **Description:** Dense coat of long yellowish-white fur. Feet heavily haired to enable it to move over ice. Ears small. Lives for up to 34 years. **Distribution:** Arctic regions of New and Old World. **Food:** Caribou, foxes, seals, walrus pups and stranded whales. **Young:** 1-4 cubs at birth. Gestation period 240 days. **Remarks:** Excellent vision and sense of smell. In winter will hunt man as it would any other animal. The largest land carnivore.

Raccoon

Raccoon

Procyon lotor

Size, etc.: 20 inches + 10-inch tail.
Description: Brownish-grey fur. Dark mask across eyes and face is notable feature of animal. Ringed tail. Lives up to 13 years. **Distribution:** Southern Canada through the United States to Central America. **Food:** Frogs, shell-fish, vegetables and seeds. **Young:** 1-7 at birth. Gestation period 60-73 days. **Remarks:** Feeds at night. Good swimmer and climber.

Weasel

Mustela nivalis

Size, etc.: 8½ inches in length (head and body). Female much smaller. **Description:** Red-brown fur. Underparts white. Lives up to 7 years. **Distribution:** North Africa, northern Asia and Europe, North America. **Food:** Mice, moles, rats, rabbits, birds, frogs. **Young:** 4-6 at birth. Gestation period 42 days. Two litters a year. **Remarks:** Generally solitary and tends to be nocturnal.

Badger

Badger

Meles meles

Size, etc.: Head and body 3 feet. **Description:** Grey fur. Black underparts and legs. White striped head. Black band from muzzle to ear. Small ears and eyes. Long claws. Lives up to 12 years. **Distribution:** Northern Asia and Europe. **Food:** Earthworms, grubs, insects, rabbits, roots. **Young:** 3-5 cubs at birth. Gestation period 168-196 days. **Remarks:** Lives in burrows and emerges at night to feed. Very playful animal.

Striped Skunk

Mephitis mephitis

Size, etc.: 2½ feet in length. **Description:** Long, black fur. White stripe along sides. Bushy tail. Short legs. Lives up to 6 years. **Distribution:** Plains and woods of U.S.A., southern Canada and Northern Mexico. **Food:** Birds, eggs, frogs, insects, snakes. **Young:** 4-7 at birth. Gestation period about 63 days. **Remarks:** When surprised it will turn its back on the attacker and squirt an evil-smelling oily liquid from the stink glands below its tail. Active at dusk and throughout the night.

Eurasian Otter

Size, etc.: Head and body 2½ feet in length. Weighs up to 25 lbs. Height 8 inches at the shoulder. **Description:** Dark brown. Underparts paler. Webbed feet. Small eyes and ears. Lives up to 11 years. **Distribution:** Africa, Asia and Europe. **Food:** Fish, birds, frogs. **Young:** 2-3 cubs at birth. Gestation period 60-63 days. **Remarks:** Excellent swimmer.

African Civet

Civettictis civetta

Size, etc.: 30 inches in length + 18-inch tail. **Description:** Thick coarse black fur with white or yellowish markings. Black feet and black and white ringed tail. **Distribution:** Africa, south of the Sahara Desert. **Food:** Small mammals, birds, fruits and berries. **Young:** 2-3 kittens at birth. **Remarks:** Civets have scent or "musk" glands which are milked from captive animals for use in the perfumery industry. The coat is also sought after by furriers.

Mongoose

African Mongoose or Ichneumon

Herpestes ichneumon

Size, etc.: 24 inches + 18-inch tail. Weighs up to 8 lbs. **Description:** Grey with buff underfur. Black-tipped tail. Live 7 to 12 years. **Distribution:** Treeless plains of Africa and southern Europe. **Food:** Birds, reptiles and other rodents. **Young:** 2-4 at birth. Gestation period about 60 days. **Remarks:** Regarded as sacred by the Ancient Egyptians. Is the largest African mongoose.

Fossa

Fossa

Cryptoprocta ferox

Size, etc.: 5 feet in length including tail. **Description:** Reddish brown, soft thick coat. Sharp, short retractile claws. Lives up to 17 years. **Distribution:** Madagascar only. **Food:** Insects and lizards, poultry, young pigs, lemurs. **Young:** 2-6 at birth. **Remarks:** Active at night.

Aardwolf

Aardwolf

Proteles cristatus

Size, etc.: 24 inches + 6-inch tail. 20 inches at the shoulder. Weighs 25-30 lbs. **Description:** Grey or buff coarse fur. Long bushy tail. Crest of hair running down back which it erects when attacked. Pointed muzzle and large erect ears. **Distribution:** Plains of eastern and southern Africa. **Food:** Insects, mainly white ants, eggs and rodents. **Young:** 2-4 at birth. **Remarks:** Not a wolf, but a close relative of the hyena. Under provocation emits an evil smelling fluid from its anal glands. Active at night.

Hyena

Spotted Hyena

Crocuta crocuta

Size, etc.: Height: 30-36 inches. Weighs 100-175 lbs. **Description:** Buff, grey or tawny fur with round black-brown spots. Massive head and sloping back. Short tail with bushy tip. Lives up to 25 years. **Distribution:** Savannahs of eastern and southern Africa. **Food:** Cattle, goats, lions' kills, sheep. Will make their own kills. **Young:** 1-2 at birth. Gestation period 90-110 days. **Remarks:** Has a distinctive high-pitched 'laughing' call during the breeding season and when excited. Nocturnal.

Lynx

European Lynx

Felis lynx

Size, etc.: 40 inches in length + 8 to 10-inch tail. Weighs up to 40 lbs. **Description:** Reddish to buffish-grey coat with dark spots. Tufted ears. Unusually broad feet. Lives up to 11 years. **Distribution:** Woods of Europe and Asia. **Food:** Rabbits, small mammals (including young deer) and birds. **Young:** 2-4 kittens at birth. Gestation period 63 days. **Remarks:** A good swimmer. Is now quite rare, especially the Spanish subspecies.

Bobcat

Bobcat

Felis rufus

Size, etc.: About 30 inches in length including short tail. **Description:** Brown fur, distinctly marked with dark spots and lines, but this can vary from region to region. **Distribution:** Southern Canada to Mexico. **Food:** Rats, mice, rabbits and snakes, but especially Snowshoe hare. **Young:** 2-4 at birth. Gestation period 40 days. **Remarks:** Most active after dark. Powerful fighters using teeth and claws.

Serval

Serval

Felis serval

Size, etc.: About 24 inches in length excluding short tail. Height 20 inches at the shoulder. **Description:** Golden-buff coloured coat profusely marked with black spots. Unusually long legs. Distinct pointed ears. Short tail. **Distribution:** Africa south of Sahara Desert. **Food:** Small mammals, birds, lizards and even insects. **Young:** 2-4 kittens at birth. Gestation period 68-74 days. **Remarks:** Travels at high speed for short distances. Rapid and skilful climber. Can catch birds by leaping up to 9 ft. in the air.

Ocelot or Painted Leopard

Felis pardalis

Size, etc.: 45-50 inches in length. Weighs about 36 lbs. **Description:** Buff or grey short-haired coat marked with bold dark spots and blotches. Ringed tail. **Distribution:** North, Central and South America. **Food:** Small animals, lizards, snakes and opossums. **Young:** 2 at birth. Gestation period 90 days. **Remarks:** Hunts mainly at night. An excellent tree climber.

Puma,

Puma, Cougar or Mountain Lion

Felis concolor

Size, etc.: 6 feet in length + 2-foot tail. Weighs 100-260 lbs. **Description:** Tawny or greyish-brown coat, whitish underparts. Lives up to 20 years. **Distribution:** Forests and plains of North, Central and South America. **Food:** Deer, small mammals and domestic cattle. **Young:** 1-4 at birth. Gestation period 90-93 days. **Remarks:** An excellent climber and swimmer. Kills with its powerful jaws.

Leopard

Leopard

Panthera pardus

Size, etc.: Average 7 feet in length, + 3 foot tail. Weighs 100-180 lbs. **Description:** Yellowish-buff coat scattered with circles or rosettes of black spots. There are variations depending on region. Lives up to 23 years. **Distribution:** Africa and Asia from Black Sea to Malaysia, including India and Ceylon. **Food:** Antelopes, monkeys, deer, small wildlife and domestic animals. **Young:** 2-5 cubs at birth. Gestation period 90-95 days. **Remarks:** Excellent tree climber, often deposits its prey in trees to enable it to feed in peace. Most active at night.

Jaguar

Jaguar

Panthera onca

Size, etc.: 4½-6 feet in length + 30-inch tail. Weighs 250-300 lbs. **Description:** Rich yellow or tawny coat, marked with a chain of black spots down the back, bordered by five rows of black rosettes running lengthwise on the sides. Its tail, limbs and head are heavily spotted and lined with black. Lives up to 22 years. **Distribution:** South-western United States, and Central and South America. **Food:** Wild pigs, capybara, monkeys, domestic cattle, alligators, turtles and fish. **Young:** 1-3 cubs at birth. Gestation period 100-110 days. **Remarks:** Is a good swimmer and an excellent climber.

Cheetah or Hunting Leopard

Acinonyx jubatus

Size, etc.: 60 inches in length + 30-inch tail. Height 30 inches at the shoulder. **Description:** Light buff-coloured coat marked all over with dark blotches. Extremely long-legged. Not a true cat as its claws are non-retractable like those of a dog. Lives up to 16 years. **Distribution:** Grassland and scrubland of Africa south of Sahara, and Asia. **Food:** Small antelopes, hares and birds. **Young:** 2-4 kittens at birth. Gestation period 84-95 days. **Remarks:** Its extremely long legs enable it to run at speeds of 45 m.p.h., or more, for bursts of about 400-500 yards.

Sea-Lion

Californian Sea-Lion

Zalophus californianus

Size, etc.: Males up to 8 feet in length. Weighs up to 600 lb. Females smaller. **Description:** Dark brown coat. Under skin there is protective layer of blubber. Head and body perfectly streamlined. · Foreflippers have five fingers, each with a rudimentary claw. Hind feet have 1st and 5th toes enlarged and clawless, the centre three toes have well developed claws. Soles of front and hind flippers naked. External ears. Lives up to 19 years. **Distribution:** Pacific coast of North America, the Galapagos Islands and Southern Sea of Japan. **Food:** Squids, shellfish and fish. **Young:** 1 at birth. Gestation period 342-365 days. **Remarks:** This is the trained seal of circuses. Gregarious throughout the year.

Walrus

Atlantic Walrus

Odobenus rosmarus

Size, etc.: Up to 11 feet. Weighs up to 3,000 lbs. Cows about one-third smaller. **Description:** Yellowish-grey, wrinkled, almost naked skin. Whiskered or moustached muzzle. Tusks (male and females) which in some males may be over 39 inches long. Flippers similar to the sea-lion. Internal ears. **Distribution:** Ice floes of Atlantic arctic. **Food:** Shellfish and sea-urchins. **Young:** 1 calf at birth. Gestation period 260-365 days.. **Remarks:** Lives in herds of 100 or more. Much reduced in numbers through hunting. Closely related sub-species in Pacific arctic.

Common Seal

Common Seal or Harbour Seal

Phoca vitulina

Size, etc.: Up to 6 feet in length. Weighs up to 300 lbs. No external ears. Hind flippers do not bend forward to support body. **Description:** White or yellow woolly pelts when newborn, moulting to grey-brown. Lives to about 19 years. **Distribution:** Northern Atlantic and Pacific. **Food:** Fish and shellfish. **Young:** 1-2 pups at birth. Gestation period 245-350 days (average 280 days). **Remarks:** Can sleep under water. Lives in loosely organized colonies.

Grey Seal

Grey or Atlantic Seal

Halichoerus grypus

Size, etc.: Up to 10 feet in length. Weighs up to 580 lbs. **Description:** Grey coat, often blotched with splashes of darker tone. Front and rear flippers are not so developed as the sea-lion's. Lives up to 18 years. **Distribution:** North Atlantic: Labrador to Novaya Zemlya and south to Channel Islands. **Food:** Fish, eels, molluscs and crustaceans. **Young:** 1 pup at birth. Gestation period almost a year. **Remarks:** Timid creature with very acute hearing.

Weddell Seal

Weddell Seal

Leptonychotes weddelli

Size, etc.: About 9 feet in length. Male weighs up to 900 lbs. Females a little larger. **Description:** Dark grey above yellow or white spots, lighter grey underneath. **Distribution:** Antarctic Ocean close to Antarctic continent. Also some subantarctic islands. **Food:** Small fish and cephalopods. **Young:** 1 pup at birth. Gestation period about 310 days. Newborn pups about half length of female. **Remarks:** Not a seal of pack ice and rarely found far from land.

Sea Elephant

Southern Sea Elephant or Elephant Seal

Mirounga leonina

Size, etc.: Up to 20 feet in length. Male weighs up to 4 tons, female smaller. Largest of all Pinnipedia. **Description:** Brownish-grey coat on coarse wrinkled skin. Underparts lighter. Inflatable snout some 8-9 inches long, on the face of the adult bull. **Distribution:** Sub antarctic waters. **Food:** Fish and cuttlefish. **Young:** 1 (rarely 2) pups at birth (usually black). Gestation period about 350 days. **Remarks:** Common name derived from large size and trunk-like proboscis. During the mating season, the proboscis is inflated when the bull roars, acting as a resonator. The roar can be heard 2 miles away.

113

Aardvark

Aardvark

Orycteropus afer

Size, etc.: 6 feet in length. 2 feet at shoulder. Weighs 140-150 lbs. **Description:** Dull brown-grey. Piglike body but with long head and tubular snout. Long pointed ears. Short strong tail. 18-inch long tongue. 4 clawed toes on fore-feet, 5 on hind-feet. Lives up to 10 years in captivity. **Distribution:** Africa south of the Sahara and Sudan. **Food:** Termites and ants. **Young:** 1 at birth. Gestation period 210 days. **Remarks:** Can dig out termite mounds at remarkable speed. Nocturnal. Excavates extensive burrows about 9-10 feet long.

Dugong

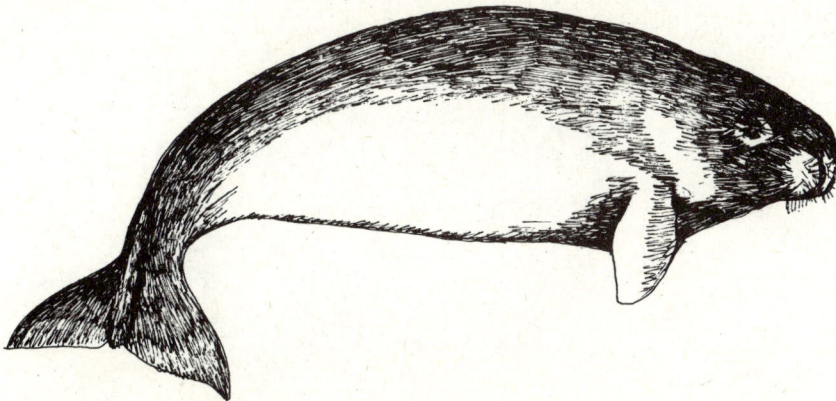

Dugong

Dugong dugon

Size, etc.: Male up to 10 feet in length, female a little shorter. Male weighs about 375 lbs. **Description:** Blue-grey. Hairless skin. Has small ears. Tail whale-like. Short tusks in male only. **Distribution:** Coastwise round Indian Ocean, Red Sea, Australia and New Guinea. **Foods:** Sea-grass and seaweeds (but not brown seaweeds). **Young:** 1 at birth. Gestation period about one year. **Remarks:** Hunted for oil and flesh. Almost exterminated now, though formerly found in very large herds.

Manatee

Manatee

*Trichechus manatus, T. inunguis,
T. senegalensis*

Size, etc.: 15 feet in length. **Description:** Grey-black almost hairless skin. Rounded tail. Eyes small. No external ears. **Distribution:** Coastal waters of Florida, the Caribbean, Northern South America and West Africa. Also the Amazon and Orinoco Rivers and the Lake Tchad drainage basin. **Food:** Underwater plants. **Young:** 1-2 calves at birth. Gestation period 152-180 days. **Remarks:** The upper lip is split down the middle and is used to pluck sea-grass.

**Przewalski Horse or
Mongolian Wild Horse**

Equus przewalskii

Size, etc.: 4½ feet at the shoulder.
Description: Brown. Black erect
mane and long-haired black tail.
No forelock. Lives for 25-30 years.
Distribution: Mongolia and Sin-
kiang. **Food:** Grasses. **Young:** 1 at
birth. **Remarks:** Only discovered in
1881. Almost extinct in the wild.

Onager

Onager

Equus hemionus onager

Size, etc.: $3\frac{1}{2}$ feet at the shoulder. **Description:** Buff with white underparts. Large ears. **Distribution:** Afghanistan, Pakistan and Persia. **Food:** Grasses. **Young:** 1 at birth. Gestation period about 1 year. **Remarks:** Large herds are led by a stallion.

American Tapir

Tapirus terrestris

Size, etc.: 3 feet at the shoulder.
Description: Dark or blackish-brown. Snout and upper lip elongated into a short flexible trunk. Body is covered with short, close hair. Four front and three hind toes. Young spotted and striped with white. Lives up to 30 years. **Distribution:** Colombia, Venezuela, Paraguay and Brazil in forests near water. **Food:** Aquatic vegetation, and leaves, buds and twigs of terrestrial plants. **Young:** 1 at birth. Gestation period 390-395 days. **Remarks:** Is inoffensive and solitary. Feeds at night.

White Rhinoceros

White Rhinoceros

Ceratotherium simum

Size, etc.: 6½ feet at the shoulder. 14½ feet in length. Weighs 3½ to 5 tons. Has 2 horns. **Description:** Slate-grey. Hump on neck. Pointed ears. Square upper lip. **Distribution:** Open plains of Zululand. **Food:** Grass. **Young:** 1 at birth. Gestation period 548-578 days. **Remarks:** Less uncertain tempered than the Black Rhino. Has poor sight but good hearing and smell.

Bush-Pig or Red River Hog

Potamochoerus porcus

Size, etc.: 25-30 inches at the shoulder. Weighs 120-180 lbs. **Description:** Long bristly reddish to blackish coat. White or black and white face patches. Two warts on face of old males. Small but sharp tusks. Long thin tail. Potentially can live 20 years. **Distribution:** Bush country and forests of Africa, south of the Sahara desert. **Food:** Bulbs, eggs, carrion, fruits, grass, roots. **Young:** 2-8 at birth. Gestation period 120-175 days. **Remarks:** Lives in groups of 6-20, sometimes more. Most active at night.

Wart Hog

Wart Hog

Phacochoerus aethiopicus

Size, etc.: 2½ feet at the shoulder. 4½ feet in length + 18-inch tail. Weighs 120-300 lbs. **Description:** Dark brown to blackish. Coarse mane on head and back. Large semi-circular upper tusks. Long, thin, tufted tail. **Distribution:** Open savannahs throughout most of Africa, especially in east and south. **Food:** Fruits, grasses, leaves and roots, occasionally carrion. **Young:** 2-4 at birth. Gestation period 171-175 days. **Remarks:** Lives in family parties. Poor sight but good hearing and smell. Usually most active in daytime.

Hippopotamus

Hippopotamus

Hippopotamus amphibius

Size, etc.: 5 feet at shoulder. Up to 14 feet in length. Weighs up to 4 tons. **Description:** Browny-grey body scantily covered with fine hair. Small ears. Large canine tusks. Short squat legs. Lives 30-46 years. **Distribution:** Large rivers and lakes of Africa: Khartoum to Zambesi. **Food:** Grasses and aquatic plants. **Young:** 1 at birth. Gestation period 233-237 days. **Remarks:** Lives in schools of 5-15 individuals. Can swim well. Generally remains in water during day, grazes on dry land at night.

Pygmy Hippopotamus

Pygmy Hippopotamus

Choeropsis liberiensis

Size, etc.: 2 feet 8 inches at the shoulder. 5 feet in length. Weighs about 500 lbs. **Description:** Slatey greenish-black, naked skin. Underparts lighter. Arched back. Relatively small head. Lives up to 40 years. **Distribution:** Forest streams and swamps of Liberia, Nigeria and Sierra Leone. **Food:** Fruits, grasses, roots, vegetables. **Young:** 1 at birth. Gestation period 201-210 days. **Remarks:** Lives singly or in pairs. Good swimmer. Nocturnal.

Dromedary

Dromedary or Arabian Camel (1 hump)

Camelus dromedarius

Size, etc.: 7½ feet high. **Description:** Yellow-brown, long, woolly coat. Lives up to 26 years. **Distribution:** Domesticated state only. Probably originally in Arabia. **Food:** Dry scrub. **Young:** 1 at birth. Gestation period 370-440 days. **Remarks:** Used as beast of burden. Cannot store water or go without water for long periods without losing weight and strength.

Chevrotain

Chevrotain or Mouse-deer

Tragulus javanicus

Size, etc.: Under 1 foot at the shoulder. 27 inches in length + 3-inch tail. **Description:** Three banded white patch on throat. Long canine teeth in upper jaw, which form tusks in the male. **Distribution:** Forests of Java, Malaya and Sumatra. **Food:** Leaves, grass, fruit and berries. **Young:** 2 at birth. Gestation period 150-160 days. **Remarks:** Shy, retiring and nocturnal. Solitary except in the mating season.

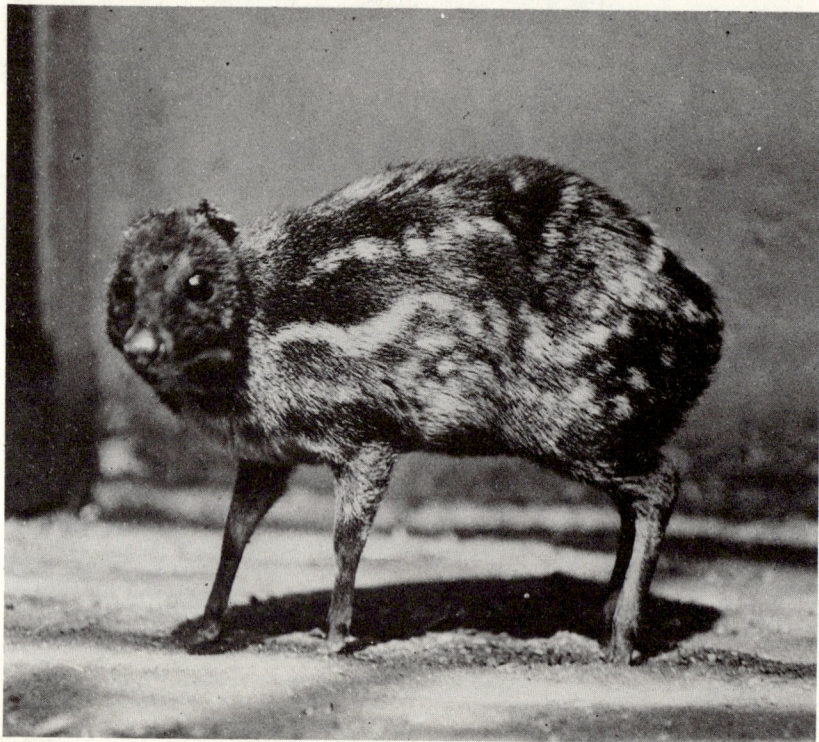

Fallow Deer

Fallow Deer

Dama dama

Size, etc.: Buck is 3 feet at the shoulder. Weighs about 100-150 lbs. **Description:** Coat, red brown with white spots in summer, greyish-fawn in winter. Underside of tail is white. Palmate antlers in male. Lives up to 15 years. **Distribution:** Originally Mediterranean region of southern Europe and Asia Minor. **Food:** Leaves. **Young:** 1 fawn at birth. Gestation period 230 days. **Remarks:** Widely introduced into parks throughout the world.

Red Deer

Red Deer

Cervus elaphus

Size, etc.: Adult stag, 4 feet at the shoulder. Weighs up to 500 lbs. **Description:** Brown-grey coat in winter. Red-brown coat in summer. Short tail. Lives to 17 or 19 years. Males bear antlers. **Distribution:** Forests and woodlands of Europe. **Food:** Leaves, grass. **Young:** 1 calf at birth. Gestation period 234-240 days. **Remarks:** Feeds at dawn and dusk. Sociable.

Reindeer

Reindeer and Caribou

Rangifer tarandus

Size, etc.: Males $3\frac{1}{2}$ feet at the shoulder. Weighs up to 650 lbs. **Description:** Dark brown or black thick coat and short tail. Peary Caribou almost white. Deeply cleft hoofs. Antlers borne by both sexes. Domesticated. Lives about 15 years. **Distribution:** Arctic regions of Europe, Asia and North America. **Food:** Grass, leaves, lichens. **Young:** 1-2 calves at birth. Gestation period 230-246 days. **Remarks:** Very gregarious. Migratory. Shy but curious.

Moose

Moose or European Elk

Alces alces

Size, etc.: Up to 7 feet 9 inches at the shoulder. Weighs up to 1,800 lbs. **Description:** Black-brown coat, becoming greyer in winter. Pendulous upper lip. Small tail. Large antlers. Lives up to 20 years. **Distribution:** Forests of Alaska, Canada and northwestern United States, Norway, Sweden, Russia and Siberia, Manchuria and Mongolia. **Food:** Shoots of saplings in winter. Aquatic vegetation in summer. **Young:** 1-3 calves at birth. Gestation period 240-250 days. **Remarks:** Senses of hearing and smell acute, vision poor.

Gerenuk

Gerenuk

Litocranius walleri

Size, etc.: Height 35-40 inches at the shoulder. Weighs 80-115 lbs. **Description:** Red-brown coat. Large eyes. Giraffe-like neck. Short, thick, ringed horns up to 17 inches (in male only). **Distribution:** Thorn-bush country of East Africa and Somaliland. **Food:** Leaves and shoots of acacia thorn bushes. **Remarks:** Lives singly or in groups of 6 or 7 females led by single male. Able to exist without water to drink.

Eland

Eland

Taurotragus oryx

Size, etc.: Height 6 feet at the shoulder. Weighs 1,300-2,000 lbs. **Description:** Greyish-fawn or tawny with white stripes on sides. Narrow pointed ears. Heavy spiral horns in both sexes, up to 45 inches long. Prominent dewlap. Long, tufted tail. Lives for 15 to 20 years. **Distribution:** Forests, grasslands and plains of Central and southern Africa. **Food:** Bulbs, fruits, leaves and shoots. **Young:** 1 calf at birth. Gestation period 255-270 days. **Remarks:** Lives in herds of 25-100. Easily tamed, and is being domesticated at Askanya-Nova in the Ukraine.

Water Buffalo

Asiatic Water Buffalo

Bubalus bubalis

Size, etc.: Height 6 feet at the shoulder. **Description:** Ash grey to black. Wide-spreading horns. Long ears. Lives about 18 years. **Distribution:** Borneo, Ceylon, China, India, Java, Sumatra. Domesticated from Egypt to the Philippines. **Food:** Water plants and lush grass. **Young:** 1-2 calves at birth. Gestation period 300-328 days. **Remarks:** Gregarious, feeding in the morning, evening and at night. Chews the cud during the daytime.

Bison

Bison or North American Buffalo

Bison bison

Size, etc.: Height $6\frac{1}{2}$ feet at the shoulder. 10 feet in length + 2-foot tail. Weighs up to 3,000 lbs. **Description:** Dark brown. Short, up-curved horns. Massive, shaggy head and thick mane. Lives up to 22 years. **Distribution:** Plains of Canada and U.S.A. **Food:** Grass. **Young:** 1 calf at birth. Gestation period 270-285 days. **Remarks:** Reduced by hunting from millions to a few thousand.

Roan Antelope

Roan Antelope

Hippotragus equinus

Size, etc.: Length 7½ feet + 2-foot tail. Height 4½-5 feet at the shoulder. Weighs 400-600 lbs. **Description:** Fawny-grey to dark grey with white underparts. Large tufted ears. Horns borne by both sexes 30-39 inches long. **Distribution:** Bush country of East, West and South Africa. **Food:** Grass. **Young:** Gestation period 270-281 days. **Remarks:** Usually travels singly or in pairs, sometimes six or so group together.

Waterbuck

Waterbuck

Kobus ellipsiprymnus

Size, etc.: 4-4½ feet at the shoulder. Weighs up to 450 lbs. **Description:** Grey or brown coat. Ringed horns up to 39 inches (in male only). Long black-tipped tail. Lives about 16 years. **Distribution:** Plains of East and West Africa. **Food:** Grass and shoots. **Young:** Usually one young at birth. Gestation period 240 days. **Remarks:** Lives in herds of 5-25.

Gemsbok

Gemsbok

Oryx gazella

Size, etc.: Height 48 inches at the shoulder. Weighs 350 lbs. **Description:** White head. Black-and-white striped face. Pale fawn-grey coat. Straight, ringed horns, 3-4 feet long. Lives about 20 years. **Distribution:** Dry plains and semi-desert of the Kalahari and S.W. Africa. **Food:** Fruits, grasses, bulbs. **Young:** 1 at birth. **Remarks:** Lives in herds of 30-40 head. Can survive without drinking.

Red Hartebeest

Red Hartebeest

Alcelaphus buselaphus

Size, etc.: Height 4 feet at the shoulder. Weighs 280 lbs. **Description:** Sandy-fawn coat. Prominent white markings on hips. Black on forehead, muzzle, shoulders. Short thick horns (in both sexes). **Distribution:** Open country of Kenya and Tanzania. **Foods:** Grass. **Young:** 1 at birth. Gestation period 214-242 days. **Remarks:** Lives in herds of 4-15 or more. Can survive without drinking.

Black Wildebeest

Black Wildebeest or White-tailed Gnu

Connochaetes gnou

Size, etc.: 6 feet in length + 2 feet tail. 4 feet at the shoulder. Weighs 550 lbs. **Description:** Dark brown to black with long-haired white tail. Massive ox-like head and horns. Lives about 16 years. **Distribution:** Game reserves of South Africa. **Food:** Grasses and succulents. **Young:** 1-2 at birth. Gestation period 240-276 days. **Remarks:** Lives in herds of 20-30. Aggressive. Almost extinct in wild.

Klipspringer

Klipspringer

Oreotragus oreotragus

Size, etc.: Overall length 3 feet 3 inches. 20 inches at the shoulder. Weighs up to 32 lbs. **Description:** Olive and yellow coat. Short straight horns (in male only). Tiny hoofs. **Distribution:** Rocky hills of eastern and southern Africa. Also in northern Nigeria. **Food:** Grasses and leaves. **Young:** 1 at birth. Gestation period 214 days. **Remarks:** Very agile. Travels in pairs or in groups of three to eight. Does not drink water regularly.

Blackbuck

Blackbuck or Indian Antelope

Antilope cervicapra

Size, etc.: Length 4 feet + 5-inch tail. Height 2 feet 8 inches at the shoulder. Weighs about 74 lbs. **Description:** Male is black-brown with white underparts. Female and young, pale brown. Twisted horns (in male only) up to 3 feet long. Lives up to 15 years. **Distribution:** Open plains of India and Pakistan. **Food:** Grasses. **Young:** 1-2 at birth. Gestation period 180 days. **Remarks:** Incredibly fast. Known to out-run attacking cheetahs.

Gazelle

Red-Fronted Gazelle

Gazella rufifrons

Size, etc.: Height 27 inches at the shoulder. Weighs 55-65 lbs. **Description:** Fawn with black and white face markings. Short ringed horns. White underparts. Lives up to 11 years. **Distribution:** Open steppes of Senegal and Northern Sudan. **Food:** Grasses and shrubs. **Remarks:** Lives singly or in herds up to 6.

Springbok

Springbok or Springbuck

Antidorcas marsupialis

Size, etc.: 4½ feet long + 7-inch tail. 2½ feet at the shoulder. Weighs 70-80 lbs. **Description:** Reddish-fawn. Underparts white. Dark band on flanks. Long fold of skin on the back. Short strong horns borne by both sexes. **Distribution:** Open plains of Angola, Kalahari Desert and South Africa. **Food:** Bulbs, grasses, leaves, roots. **Young:** 1 at birth. Gestation period about 171 days. **Remarks:** Leaps 8-10 feet into air with arched back and fully extended legs. Undertakes long migrations. Is the national emblem of South Africa.

Chamois

Chamois or Gems

Rupicapra rupicapra

Size, etc.: $3\frac{1}{2}$ feet long + $1\frac{1}{2}$-inch tail. Height $2\frac{1}{2}$ feet at the shoulder. Weighs up to 90 lbs. **Description:** Coat tawny and short in summer, browny-black and long in winter. Sharp-pointed backward-curving horns. Lives about 22 years. **Distribution:** High forest belts of the European Alps, Asia Minor and Carpathians. **Food:** Grass, leaves and shoots. **Young:** 1-3 kids. Gestation period about 180 days. **Remarks:** Remarkable climber. Can jump 20 feet. Lives in herds of 15 to 30 females and young. Male solitary except in autumn mating season.

Musk-Ox

Ovibos moschatus

Size, etc.: Length 8 feet. $4\frac{1}{2}$ feet at the shoulder. Weighs up to 800 lbs. **Description:** Brown, shaggy coat of guardhairs and light brown underfur. Black underparts. Recurved horns. Broad hoofs. Lives about 20 years. **Distribution:** Northern Canada, Greenland. Introduced to Norway. **Food:** Grass, moss and arctic willow. **Young:** 1 calf at birth. Gestation period 270 days. **Remarks:** Lives in small herds. For defence, a herd forms an outward facing circle with young in the centre.

Ibex

Ibex

Capra ibex

Size, etc.: Height up to 3 feet at the shoulder. 4½ feet long + 5-inch tail. Weighs up to 200 lbs. **Description:** Grey-brown. Upward-curving heavily ridged horns, up to 44 inches long. Bearded. **Distribution:** European Alps. **Food:** Vegetation. **Young:** 1-2 kids at birth. Gestation period 240-280 days. **Remarks:** Also found in Spain and Portugal and in Asia and North Africa.

Mouflon

Size, etc.: 3½ feet long + 4-inch tail. 27 inches at the shoulder. **Description:** Red-brown. White underparts. Large backward-curving horns. **Distribution:** Corsica and Sardinia. Introduced widely in Europe. **Food:** Grasses, leaves and flowers. **Young:** 1-3 young at birth. Gestation period 150-180 days. **Remarks:** The only European wild sheep.

THE CLASSIFICATION OF LIVING MAMMALS

CLASS MAMMALIA

SUB-CLASS PROTOTHERIA
ORDER MONOTREMATA

		SPECIES
Family Tachyglossidae	Echidnas	5
Family Ornithorhynchidae	Duck-billed platypus	1

SUB-CLASS METATHERIA
ORDER MARSUPIALIA

Family Didelphidae	Opossums	65
Family Dasyuridae	Native cats	45
Family Myrmecobiidae	Numbat	1
Family Notoryctidae	Marsupial moles	2
Family Peramelidae	Bandicoots	19
Family Caenolestidae	Rat opossums	7
Family Phalangeridae	Possums	45
Family Vombatidae	Wombats	2
Family Macropodidae	Kangaroos	52

SUB-CLASS EUTHERIA
ORDER INSECTIVORA

Family Solenodontidae	Solenodons	2
Family Tenrecidae	Tenrecs	20
Family Potamogalidae	Otter shrews	3
Family Chrysochloridae	Golden moles	20
Family Erinaceidae	Hedgehogs	15
Family Soricidae	Shrews	200+
Family Talpidae	Moles	19
Family Macroscelididae	Elephant shrews	14

ORDER DERMOPTERA

Family Cynocephalidae	Colugos	2

ORDER CHIROPTERA
SUB-ORDER MEGACHIROPTERA

		SPECIES
Family Pteropodidae	Fruit bats	130

SUB-ORDER MICROCHIROPTERA

Family Rhinopomatidae	Mouse-tailed bats	4
Family Emballonuridae	Sheath-tailed bats	50
Family Noctilionidae	Bulldog bats	2
Family Nycteridae	Slit-faced bats	10
Family Megadermatidae	False vampires	5
Family Rhinolophidae	Horseshoe bats	50
Family Hipposideridae	Leaf-nosed bats	100
Family Phyllostomatidae	American leaf-nosed bats	100
Family Desmodontidae	Vampires	3
Family Natilidae	Funnel-eared bats	115
Family Furipteridae	Smoky bats	2
Family Thyropteridae	Disc-winged bats	2
Family Myzopodidae		1
Family Vespertilionidae	Insectivorous bats	275
Family Mystacinidae		1
Family Molossidae	Mastiff bats	80

ORDER PRIMATES
SUB-ORDER PROSIMII

Family Tupaiidae	Tree shrews	20
Family Lemuridae	Lemurs	15
Family Indriidae	Indri and Sifakas	4
Family Daubentoniidae	Aye-aye	1
Family Lorisidae	Lorises and Bushbabies	11
Family Tarsiidae	Tarsiers	3

SUB-ORDER ANTHROPOIDEA		SPECIES
Family Callitrichidae	Marmosets and Tamarins	21
Family Cebidae	Cebid monkeys	26
Family Cercopithecidae	Old World monkeys	60
Family Pongidae	Apes	9
Family Hominidae	Man	1

ORDER CETACEA
SUB-ORDER ODONTOCETI

Family Platanistidae	River dolphins	4
Family Ziphiidae	Beaked whales	15
Family Physeteridae	Sperm whales	2
Family Monodontidae	White whale and Narwhal	2
Family Delphinidae	Dolphins and Porpoises	50

SUB-ORDER MYSTICETI

Family Eschrichtiidae	Grey whale	1
Family Balaenopteridae	Rorquals	6
Family Balaenidae	Right whales	3

ORDER EDENTATA

Family Myrmecophagidae	Anteaters	3
Family Bradypodidae	Sloths	5
Family Dasypodidae	Armadillos	20

ORDER PHOLIDOTA

Family Manidae	Pangolins	7

ORDER LAGOMORPHA

Family Ochotonidae	Pikas	14
Family Leporidae	Hares and Rabbits	50

ORDER RODENTIA
SUB-ORDER SCIURIDAE

Family Aplodontidae	Sewellel	1
Family Sciuridae	Squirrels	250

Family Geomyidae	Pocket gophers	30
Family Heteromyidae	Kangaroo rats	70
Family Castoridae	Beaver	1
Family Anomaluridae	Scaly-tailed squirrels	9
Family Pedetidae	Springhaas	1

SUB-ORDER MYOMORPHA

Family Cricetidae	Hamsters	570
Family Spalacidae	Mediterranean mole-rats	3
Family Rhizomyidae	Bamboo rats	18
Family Muridae	Rats and Mice	500
Family Gliridae	Dormice	10
Family Platacanthomyidae	Spiny dormouse	2
Family Seleviniidae	Desert dormouse	1
Family Zapodidae	Birch mice	11
Family Dipodidae	Jerdoas	25

SUB-ORDER HYSTRICOMORPHA

Family Hystricidae	Old World Porcupines	20
Family Erethizontidae	New World Porcupines	23
Family Caviidae	Guinea pigs	23
Family Hydrochoeridae	Capybara	1
Family Dinomyidae	Pacarana	1
Family Dasyproctidae	Agoutis	30
Family Chinchillidae	Chinchillas	6
Family Capromyidae	Coypus and Hutias	10
Family Octodontidae	Octodonts	8
Family Ctenomyidae	Tuco-tucos	26
Family Abrocomidae	Chinchilla rats	2
Family Echimyidae	Spiny rats	75
Family Thryonomidae	Cane rats	6
Family Petromyidae	Rock rat	1
Family Bathyergidae	African mole-rats	50
Family Ctenodactylidae	Gundis	8

ORDER CARNIVORA		
SUB-ORDER AELUROIDEA		SPECIES
Family Felidae	**Cats**	34
Family Hyaenidae	**Hyenas and Aardwolf**	4
Family Viverridae	**Civets and Mongooses**	75
SUB-ORDER ARCTOIDEA		
Family Canidae	**Dogs**	37
Family Mustelidae	**Weasels**	70
Family Ursidae	**Bears**	7
Family Procyonidae	**Raccoons and Pandas**	18
ORDER PINNIPEDIA		
Family Odobaenidae	**Walrus**	1
Family Otariidae	**Sea lions**	13
Family Phocidae	**True seals**	18
ORDER PROBOSCIDEA		
Family Elephantidae	**Elephants**	2
ORDER HYRACOIDEA		
Family Procaviidae	**Hyraxes or Dassies**	6
ORDER SIRENIA		
Family Trichechidae	**Manatees**	3
Family Dugongidae	**Dugong**	1

ORDER TUBULIDENTATA		SPECIES
Family Orycteropodidae	**Aardvark**	1
ORDER PERISSODACTYLA		
SUB-ORDER CERATOMORPHA		
Family Tapiridae	**Tapirs**	4
Family Rhinocerotidae	**Rhinoceroses**	5
SUB-ORDER HIPPOMORPHA		
Family Equidae	**Horses and Zebras**	6
ORDER ARTIODACTYLA		
SUB-ORDER SUIFORMES		
Family Suidae	**Pigs**	8
Family Tayassuidae	**Peccaries**	2
Family Hippopotamidae	**Hippopotamuses**	2
SUB-ORDER TYLOPODA		
Family Camelidae	**Camels**	3
SUB-ORDER RUMINANTIA		
Family Tragulidae	**Chevrotains**	4
Family Cervidae	**Deer**	40
Family Bovidae	**Cattle and Antelopes**	100
Family Antilocapridae	**Pronghorn**	1
Family Giraffidae	**Giraffes**	2

Index